NIGHT SKY

NIGHT SKY

STARGAZING WITH THE NAKED EYE

ROBERT HARVEY

amber
BOOKS

Reprinted 2020

First published in 2019

Copyright © 2019 Amber Books Ltd

Published by
Amber Books Ltd
United House
North Road
London
N7 9DP
United Kingdom
www.amberbooks.co.uk
Instagram: amberbooksltd
Facebook: amberbooks
Twitter: @amberbooks
Pinterest: amberbooksltd

Project Editor: Sarah Uttridge
Designers: Keren Harragan
Picture Research: Terry Forshaw

ISBN: 978-1-78274-918-9

Printed in China

Contents

Introduction

More than half the world's human population lives in cities. Apart from the Moon and a few bright stars, it is likely that most people rarely see the night sky.

For city-dwellers, to travel away from the glare of fluorescent lights to somewhere that the sky is truly dark is a revelation. Until the invention of electric light at the end of the nineteenth century, the night sky was universally accessible. In just over a century, most of the population has lost a view that for thousands of years, humans took for granted.

This book depicts the glory of the night sky from every inhabited continent. All the astronomical objects illustrated can be seen with the unaided eye, and none of the pictures were taken with a telescope (except for the nebula on page 10). These views are available to everyone. They are part of the common heritage of humankind.

ABOVE:
A partial solar eclipse on 4th December 2012, Broken Bay, Sydney, Australia
OPPOSITE:
The Milky Way, Red Rock Canyon, California, USA

The photographs in this book were taken with regular cameras and lenses. In order to photograph the night sky, you need a digital camera with manual control over exposure. You need to be able to set the aperture and shutter speed independently and to select a high ISO. A digital single lens reflex (SLR) camera with a full-frame sensor is ideal. A compact camera or phone camera won't do.

GETTING THE PERFECT PHOTO

A sturdy tripod is essential. It takes many seconds to collect enough light for an image of the night sky, during which time your camera must be completely steady.

Find a good dark sky, away from city lights. Areas of remote countryside, mountains, coasts and deserts are usually ideal. Look for a strong foreground subject, something that stands against the sky such as peaks, rocks, towers or trees. Be sure to take a dependable torch to find your way there and back, a head torch to see what you are doing and perhaps a hand torch to light your foreground subject.

Finally, do your research! Know what you are planning to photograph, when it will be visible and how bright it will be. Check the weather forecast and phase of the Moon – a little moonlight can illuminate a foreground subject but a Full Moon will overwhelm the stars.

ABOVE:
Camera Equipment
There are a variety of techniques that can be used to create impressive photographs of the night sky. However, a solid tripod is essential for astrophotography as it usually takes 20 to 30 seconds to collect enough light to compose a correctly exposed picture of the night sky.

WHAT YOU CAN SEE IN THE NIGHT SKY WITH THE NAKED EYE

- The Sun and Moon
- Aurora (Northern and Southern Lights)
- Meteors
- The International Space Station
- Five planets: Mercury, Venus, Mars, Jupiter and Saturn
- Bright comets (when they appear in the sky)
- About 2000 stars in a very dark sky
- The Milky Way
- Large and Small Magellanic Clouds (Southern Hemisphere)
- Andromeda Galaxy (Northern Hemisphere)

HOW TO TAKE PHOTOS: DO'S AND DON'TS

When photographing the night sky, you need to collect as much light as possible. So set your lens to its widest aperture (lowest F/number). For stars and planets, use a shutter speed of 20 to 30 seconds and an ISO of between 400 and 1600. For the Milky Way, recommended settings are 20 seconds with an aperture of F/2 and an ISO of 3200.

Use a remote release or self-timer, so you are not touching your camera while the shutter is open.

When venturing into remote areas, check out the location in daylight first and go with someone else for safety. Always take a back-up torch and a map to be sure of finding your way home.

LEFT:

Los Angeles Downtown and surrounding area at sunset, California, USA

The brightly lit centre of Los Angeles creates urban sky glow which, in combination with an urban heat haze, makes the night sky virtually invisible. Many urban-dwellers have never seen the Milky Way and must travel long distances to escape the effects of light pollution.

BOTTOM LEFT:

Night Sky over Karoo National Park, South Africa

Karoo National Park encompasses a huge tract of semi-desert in South Africa's Western Cape. Far from the densely populated coast, the night sky is studded with stars, of which about 2000 are typically visible to the unaided eye. Orion and Taurus, with its Pleiades star cluster, are clearly visible.

NEXT PAGE:

Hubble Telescope view of the Eagle Nebula

There are an estimated 250 billion stars in our galaxy, which is one of at least 100 billion galaxies in the Universe. With the naked eye we can see only a fraction of these, but orbiting above the Earth, Nasa's Hubble Space Telescope takes images that cannot be captured from the ground. This view of the Eagle Nebula shows an area of active star formation about 7000 light years from Earth.

Europe

Much of Europe is densely populated. In many people's lives, the glare of urban lights usually seems close at hand. But for most, an hour or two of travel is sufficient to reach relatively dark skies and uncluttered horizons.

The coasts and mountains of Europe are spectacular locations for star-gazing. The Alps, whose soaring peaks stand high against the night sky, would be high on many people's list. Equally enticing is a reflection of the stars in an alpine lake. For the greatest celestial light show, the Aurora, the Nordic countries offer some of the best viewing locations on Earth. The spectacular geography of Norway and Iceland makes stunning foregrounds to the Northern Lights.

Europe has seen many advances in astronomy and our understanding of our place in the Universe. In sixteenth-century Poland, Nicolaus Copernicus proposed that the Earth orbits the Sun. In seventeenth-century Italy, Galileo Galilei invented the telescope and discovered Jupiter's moons, providing powerful support for Copernicus. In eighteenth-century England, William Herschel discovered Uranus, a new planet in our solar system. In nineteenth century France, Urbain Le Verrier used Newton's Laws of Physics to correctly predict the existence and location of Neptune.

Astronomy was practised at Stonehenge more than 4000 years ago. Today, such sites provide a link with our past and a place to seek experiences of the night sky that have something in common with those of our ancestors.

OPPOSITE:
Supermoon rising over Glastonbury Tor, Somerset, England
A 'supermoon' occurs when a Full Moon coincides with the Moon's closest approach to Earth in its orbit (perigee). The Moon then appears about 10 per cent bigger and 30 per cent brighter than when it occurs at its furthest point from Earth (apogee). The Full Moon of 16 November 2016 was the closest to Earth for 68 years and will not be surpassed until 2034. Viewed from a distance of more than 2km (1.2 miles), the Full Moon appeared as big as St Michael's Tower in Glastonbury.

LEFT:

Aurora Borealis over Aldeyjarfoss Waterfall, Iceland

The Aurora Borealis, or Northern Lights, occur when charged particles from the Sun strike the outermost reaches of Earth's atmosphere. These cause electrons in atmospheric gases to be temporarily boosted into higher orbits, referred to as being 'excited'. When the electrons drop back to their usual state, energy is released in the form of light of a characteristic wavelength. The green colour most commonly observed in aurorae results from oxygen atoms becoming excited.

RIGHT:

Watching the Northern Lights, Jokulsarlon, Iceland

As charged solar particles approach Earth they become caught up in our planet's magnetic field, which directs them towards the polar regions. This is why the Northern Lights are most commonly seen around the Arctic Circle. Intense aurorae can appear as shimmering curtains of light, rapidly changing in shape and moving across the sky. This observer is holding a camera in his hand but he is not taking photographs – he is just experiencing the unfolding spectacle.

ABOVE AND RIGHT:

Moon, Venus, Pleiades and Hyades over Kiruna, Sweden

Taiga, or boreal forest, is brightly lit by a Crescent Moon on a dark winter night in northern Lapland. Venus lies to the left of the Moon, just visible between tree branches. At the top left are two star clusters: the Pleiades and the Hyades, both in the constellation of Taurus. The Pleiades are hot young stars formed within the last 100 million years, located 444 light years from Earth. The Hyades are older stars, forming a wider 'V' shaped cluster, at a distance of 153 light years (see inset above).

OPPOSITE:

Aurora Borealis over Kirkjufell Mountain, Iceland

Kirkjufell 'Church Mountain' is a beautiful cone-shaped mountain on the north coast of Iceland's Snæfellsnes Peninsula. In winter, the waterfall at the base of the mountain freezes while the Northern Lights dance in the sky above.

Snow Moon, Lofoten Islands, Norway
Each Full Moon of the year has its own name. February's Full
Moon is known as the 'Snow Moon', seen here rising above the
snow-clad mountains of Hinnoya in Norway's Lofoten Islands.
At 3.30 in the afternoon, the Sun has already set but a little
sunlight is reflected from ice particles in Earth's atmosphere
to give alpenglow on the peaks.

Aurora Borealis over Sommarøy Bridge, Norway
Sommarøy Bridge connects the islands of Kvaløya and
Sommarøy in Tromsø Municipality, north of the Arctic Circle.
Sweeping across the northern sky, this aurora is bright enough
to be seen against a backdrop of city lights.

Dance of the Northern Lights, Lofoten Islands, Norway
Haukland Beach on the island of Vestvagoya in Norway's
Lofoten Islands occupies a spectacular location between steep-
sided coastal mountains, with rock ridges running towards
the sea. Most aurorae appear white to the human eye but the
subtle shapes and colours of this aurora have been picked up
by the camera.

Intense aurora, Lofoten Islands, Norway

The Lofoten Islands are a popular Northern Lights location owing to their spectacular mountain landscapes and excellent road access. Photographs taken simultaneously from different points on the ground have shown that most aurora occur between 90 and 150km (60–90 miles) above the surface of the Earth, though a few are formed as high as 1000km (600 miles).

22

LEFT:

Road to the Northern Lights, Norway

The brightest aurorae can appear like curtains descending from heaven. Each curtain comprises many parallel rays, aligned by Earth's magnetic field, sometimes resembling folds in a sheet of fabric. Here, the auroral display composed of both green, emitted by excited oxygen atoms, and purple, from molecular nitrogen struck by solar particles.

RIGHT:

Venus over Kald Fjord, Kvaloya, Norway

Venus, our nearest planetary neighbour, is the third brightest object in the night sky, up to 16 times brighter than the most luminous star. This picture was taken by the light of an almost Full Moon, illuminating the mountains of Kvaloya Island and turning the sky blue, swamping the stars. Nearing eastern elongation from the Sun as viewed from Earth, Venus dominates the western sky.

OPPOSITE:
Milky Way above mountains, Norway
The Milky Way is the galaxy in which we live. It is a flattened spiral disk, with our solar system in one of the outer arms. When we look along the plane of that disk, we see a faint band of light made up of millions of stars that are too distant to make out individually. Seeing the Milky Way requires a dark sky away from light pollution and a moonless night.

ABOVE:
Venus and Crescent Moon, Lapland, Finland
Viewed just before dawn, the Crescent Moon appears close to Venus in the eastern sky. The lunar crescent, lit directly by the Sun, is overexposed to show the faintly illuminated night side of the Moon. We can see this thanks to Earthshine – sunlight reflected from the day time side of Earth onto the lunar surface.

LEFT:
Milky Way over Tatra Mountains, Poland
In spring, the bright core of the Milky Way rises just before dawn. In this spectacular panorama, made high in the rugged Tatra Mountains, we see the whole arc of the Milky Way stretching across the eastern sky. On the right-hand side, we are looking towards the galactic centre, whereas the left-hand side shows the fainter outer reaches of our galaxy.

ABOVE:
Milky Way over the Carpathians, Ukraine
The faint band of the Milky Way arches over the Carpathian Mountains in winter. Recent snow has left the trees coated, as if by icing sugar. Clear mountain air and an absence of light pollution has enabled far more stars to be recorded by the camera than can be seen with the human eye.

Russia, Lake Elton salt lake

Lake Elton is a vast salt lake, the biggest mineral lake in Europe and is located near Russia's border with Khazakstan. Made just before dawn in May, this 180-degree panorama shows the whole arch of the Milky Way from the galactic core on the right to the outlying regions of our galaxy on the left. The constellation of Scorpius, with its bright star Antares, is on the extreme right of the image. Mars, the Red Planet, is the brightest object in the sky, just above Antares. Saturn, of similar brightness to Antares, lies to its left.

Star trails above Enisala Fortress, Dobrogea, Romania
As Earth rotates, all the stars in the night sky appear to circle the North Pole star, which remains almost stationary in the sky. While this effect of Earth's rotation is imperceptible to our eyes, the camera can record it as star trails. This image is built up from many individual frames taken over a period of about an hour, which were then combined in processing.

ABOVE:
Perseid Meteors above the Beglik dam, Bulgaria
Each year on around 12 August, Earth enters a region of space containing a stream of debris left by Comet Swift–Tuttle. Most of this debris comprises particles no bigger than a grain of sand, but they are travelling at tremendous speed relative to Earth. Colliding with our upper atmosphere, these meteoroids burn up in a brief trail of light. The Perseids take their name from the constellation of Perseus, from which they appear to radiate.

LEFT:

Fairy chimneys at night with stars in the sky, Cappadocia, Turkey

The fairy chimneys of Cappadocia are natural spires of rock created by weathering in a semi-arid basin. They are formed as a result of softer rock overlain by harder rock, which protects each column from erosion. Many of Cappadocia's chimneys have been excavated and used as dwellings. In this image taken by moonlight, the constellation of Taurus can be seen above the tallest fairy chimney.

RIGHT:

Milky Way, Mount Olympus, Greece

Mount Olympus is the highest mountain in Greece, with a summit elevation of over 2918m (9573ft). Framed by trees on the mountain's flanks, the core of the Milky Way was visible low in the southern sky during August 2018. Mars, just two weeks past its closest approach to Earth for 15 years, can be seen between the branches of the dead tree on the left. This photograph was taken at an elevation of over 2000m (6500ft), where exceptional clarity to the air and a long exposure have enabled the Milky Way to be recorded in great detail.

Braies Lake, South Tyrol, Italy

Braies Lake is the largest natural lake in the Dolomites, and is known for its natural beauty and spectacular setting. At an elevation of 1496m (4910ft), Braise is turquoise in summer and frozen in winter. This image was taken at the very end of winter looking south-east, as the core of the Milky Way starts to rise between the mountains overlooking Braies Lake.

Orion over Colma di Sormano, Como, Italy
Named after a great hunter in Greek mythology, Orion is one of the most instantly recognizable constellations. It has seven bright stars (magnitude 0, 1 or 2), of which two form the hunter's shoulders, two represent his knees and three his belt.

OPPOSITE FAR LEFT:

Milky Way over Sibillini Mountains National Park, Le Marche, Italy
An atmospheric inversion layer at night can cause fog to form at lower elevations, with clear skies above. This creates excellent viewing conditions for the Milky Way, since fog screens light pollution from the valleys.

RIGHT:

Milky Way on Summer night, Dolomites, Italy
Looking away from the Galactic centre, this view captures the fainter northern part of the Milky Way, framed by some thin cloud.

Milky Way, Bardenas Reales, Spain

Bárdenas Reales is an area of semi-desert badlands in northern Spain. Wind and water erosion have created canyons, mesas and cabezos (isolated hills). There are few settlements in the region so the sky is sufficiently dark for the Milky Way to be clearly visible. Both the sky near the horizon and the nearby landscape are illuminated by stray light from distant towns.

OPPOSITE:

Night Sky over the Matterhorn, the Alps, Switzerland

The distinctive shape of the Matterhorn results from multiple glaciers diverging and carving out four faces to the north, south, east and west, separated by ridges. Using a long telephoto, this view of the forbidding summit has captured many fainter stars in the background.

LEFT:

Orion over Ves chapel, Vosges, France

Chapel des Vés was built in 1863 by local farmers and reconstructed in its original style in 2000. The constellation of Orion, high in the northern hemisphere sky during winter evenings, is right above the chapel. Follow a line formed by the three stars in Orion's belt to the left to find Sirius, the brightest star in the night sky.

BELOW:

Stars over the Eiffel Tower, Paris, France

A digital camera can record stars even over one of the largest cities in Europe. Completed in 1889, the Eiffel Tower is a cultural icon of France. Its height is 324m (1064ft), which for 41 years made it the tallest building the world. The faint band of the Milky Way can be seen to its right.

Starry sky over Mont Blanc Range, Lac de Chesery, France/Italy/Switzerland
It is not often that atmospheric conditions are clear and still enough to see perfect reflections of stars in an alpine lake. In this image, the faint band of the Milky Way is seen above the massif of Mont Blanc, at 4809m (15,778ft) the highest mountain in the Alps. The mountain itself lies along the border of France and Italy, whilst Lac de Cheserey, in which it is reflected in this image, is in Switzerland.

FAR LEFT:

Cassiopeia and Milky Way reflected in Garreg-ddu reservoir, Elan Valley, Wales

The Elan Valley is an International Dark Sky Park and the least light-polluted area in Wales. In this image, we see the northern part of the Milky Way arching over a Victorian water intake structure, with the W-shaped constellation of Cassiopeia above it and reflected in the still water.

NEAR LEFT:

Eclipsed Moon and Mars, Munich, Germany

During a total lunar eclipse, the Moon lies within Earth's shadow but is faintly illuminated by indirect sunlight refracted through Earth's atmosphere, turning it deep red. On 27 July 2018, a total lunar eclipse coincided with Mars at Opposition to Earth, meaning both appeared close together in the sky, rising at sunset. This image shows the blood-red Moon and the Red Planet above Munich airport.

OPPOSITE:

Geminid Meteor over Stonehenge, Wiltshire, England

The Geminids are the brightest annual meteor shower, peaking around 13 December each year. The meteor shower is associated with an asteroid, 3200 Phaethon, which has left a trail of debris along its orbit.

International Space Station over Severn Bridge, England/Wales

The International Space Station is the second brightest object in the night sky after the Moon. It makes a complete orbit of Earth every 92 minutes and, as seen by an observer on the ground, crosses the sky from west to east in about 10 minutes. A sequence of 10-second exposures has captured the trail of the space station over the Severn Bridge; terrestrial and celestial transport in a single image.

LEFT AND ABOVE:

Venus, Mars and Jupiter in conjunction, Isle of Portland, England

Venus is the brightest of the three planets in this image, seen just above the horizon. Jupiter is the largest planet in the solar system but it is around 15 times further away than Venus, so appears about one-fifth as bright. Mars is the dimmest of the three, because at the time of the conjunction it was near its furthest point from Earth in its orbit.

RIGHT:

Ursa Major over Lacock Abbey, Wiltshire, England

Above the moonlit Abbey lies the easily recognized constellation of Ursa Major, whose seven bright stars are referred to as 'The Plough' in Britain. Follow the curve of The Plough's 'handle' to the left to find Arcturus, the brightest star in the northern celestial hemisphere.

OPPOSITE LEFT:

Milky Way over St Michael's Mount, Cornwall, England

St Michael's Mount is a tidal island in Mount's Bay, accessible from the Cornish mainland at low tide by a causeway. In August, the core of the Milky Way comes into view, low on the southern horizon around midnight and perfectly aligned as seen from the causeway. Although the buildings on St Michael's Mount are lit, there is no light pollution beyond the island.

OPPOSITE RIGHT:

Moon and Venus, St Michael's Mount, Cornwall, England

This view of St Michael's Mount was taken from the landward end of the causeway just after sunset in winter. Venus and the crescent Moon are low in the southern sky and will set soon after the Sun.

RIGHT:

Super Blue Moon Rise, Wiltshire, England

A 'Blue Moon' is the name given to a second Full Moon in a month. This one occurred close to Earth in the lunar orbit, so it was also a supermoon. Photographed rising behind a tree on the Marlborough Downs, we can see that a Blue Moon is the same colour as any other Full Moon!

ABOVE AND RIGHT:

Orion and Pleiades over Martinsell Hill, Wiltshire, England
Moonlit on a February evening, this Scots Pine stands sentinel
on a windy hilltop, its branches softened by motion blur
during a 20-second exposure. To the right is the most obvious
star cluster in our night sky, the Pleiades (also seen in the
inset above), whilst to the left of the tree is the constellation
of Orion. Six of the Pleiades can be seen with the naked eye,
whilst a telescope reveals many more.

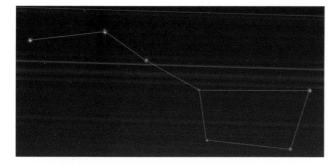

ABOVE AND OPPOSITE:

The Great Bear over Avebury, Wiltshire, England
These two standing stones in Avebury's northern inner circle
are illuminated by gibbous moonlight, whilst above them are
the bright stars of Ursa Major, the Great Bear. Interestingly,
the identification of this constellation as a celestial bear is
shared by widely separated civilizations across the northern
hemisphere, suggesting the idea originated as an oral tradition
thousands of years ago in human pre-history.

LEFT AND ABOVE:

Regulus, Castor, Pollux and Procyon, Isle of Skye, Scotland

The Old Man of Storr is one of several weirdly shaped rock formations created by landslips on Skye's Trotternish peninsula. Four first magnitude stars are visible in this image: Regulus (in the constellation of Leo) on the left, Castor and Pollux (in Gemini) on the right and Procyon (in Canis Minor) in the rock cleft. Castor and Pollux are also shown in the inset above.

OPPOSITE:

Moonrise on Lundy, Devon, England

The Old Light on Lundy was built in 1819 to warn mariners that they were approaching the island. It was replaced by two lower lighthouses and now serves as a holiday cottage. June's Full Moon, the Strawberry Moon, rose about 20 minutes after sunset, when there was just enough fading light in the sky to illuminate the old lighthouse in balance with the Moon.

Fireball over Clatteringshaws Loch, Dumfries and Galloway, Scotland

A fireball is a meteor that is brighter than any of the planets. On 21 September 2012, this spectacular fireball broke into fragments as it entered the atmosphere over southern Scotland, resulting in three parallel trails across the sky. Fireballs as bright as this, which are of rocky or metallic composition, may survive their passage through Earth's atmosphere and reach the ground as meteorites.

North America

The National Parks of the United States and Canada are wonderful places to experience the night sky. Sandstone arches, granite mountains and crystal clear lakes are far from city lights. On a moonless night, the main illumination is starlight, along with a little reflected sunlight from space. It is on such nights that we can see the Milky Way – a pale ribbon of light across the sky. Our galaxy is a flattened spiral disk, and when we see the Milky Way, we are looking along the plane of that disk. We see billions of stars, too distant and numerous to make out individually, as a faint band of light.

Observations made by American astronomer Edwin Hubble in the early twentieth century discovered that many faint objects, previously thought to be clouds of dust and gas, are actually galaxies beyond the Milk Way.

This demonstrated that contrary to belief at the time, the Milky Way is not the only galaxy in the Universe. In fact, the Universe contains countless billions of galaxies. In 1990, NASA launched the Hubble Space Telescope, which continues to peer deep into space. Its many discoveries include determining the age of the universe to be 13.7 billon years.

Long before telescopes were invented, Native Americans described what they saw in the night sky, developing elaborate myths to explain the Aurora Borealis. Like the Europeans, they saw the constellation Ursa Major as a great bear, suggesting this interpretation originates deep in prehistory before people reached the Americas. Native Americans gave us the names still used for each month's Full Moon such as January's Wolf Moon and August's Sturgeon Moon.

OPPOSITE:
Milky Way over Lake Louise, Banff National Park, Canada
Lake Louise is one of the best known and most beautiful locations in the Canadian Rockies. It receives a great many visitors in August, but around midnight on a moonless night, the lakeshore is deserted. Looking south west, the bright core of our Milky Way galaxy directly over Victoria Glacier makes a stunning sight.

ABOVE:
Aurora Borealis, Greenland
Greenland, the world's largest island, lies right under the auroral belt. This region, around which the Northern Lights are most often seen, is centred not on the geographical North Pole but around Earth's magnetic North Pole, currently located north of the Canadian Arctic. Photographed during the long Arctic night, this iceberg reflects the mysterious green light of the aurora.

OPPOSITE:
Aurora Borealis, Greenland
Reflected in an icy sea, this aurora appears like the wavy hem of a curtain descending from space. The distinct shape results from charged particles in the solar wind being aligned by Earth's magnetic field as they are funnelled towards the north magnetic pole, whilst the colour comes from interaction with oxygen atoms in the Earth's outer atmosphere.

62

Northern Lights over Tasiilaq, Greenland
With a population of just over 2000, the Inuit village of Tasiilaq is Greenland's seventh largest town. Lying just 106km (66 miles) south of the Arctic Circle, long winter nights are frequently illuminated by the dancing lights of the Aurora Borealis.

Milky Way over Kathleen Lake and the St Elias Mountains, Kluane National Park, Yukon, Canada

Kluane National Park in Canada's Yukon Territory is a vast wilderness of ice fields, forests and towering peaks, including Canada's highest mountain, Mount Logan. At a latitude of 60 degrees, the core of the Milky Way never rises this far north but with few inhabitants, the skies are dark enough to see the outlying regions of our galaxy.

Milky Way, Writing-on-Stone Provincial Park, Alberta, Canada

Looking due south, a moonless night at the end of July shows the full glory of our galaxy over sandstone formations in Writing-on-Stone Provincial Park. Located in southern Alberta, this park contains the greatest concentration of rock art on the North American Great Plains, with over 50 petroglyph sites and thousands of artworks.

Northern Lights over North Saskatchewan Landing School, Kyle, Canada
The multi-hued pastel colours of this swirling aurora result from the interaction of charged particles in the solar wind with different gases in the rarified outer reaches of Earth's atmosphere. Green is emitted by excited electrons in atomic oxygen and purple by molecular nitrogen, while yellow results from the additive mixing of these colours.

LEFT:

The Milky Way over the Oxtongue Rapids, Muskoka, Canada
Early fall colours around the Oxtongue River are illuminated by moonlight. The faint band of the Milky Way can be seen in the sky above.

RIGHT:

Supermoon rising over Toronto, Ontario, Canada, 13 November 2016
A supermoon is a Full Moon that occurs when the Moon is close to Earth in its orbit. As a result, the Moon appears slightly larger in our sky than at other times. Many people have noticed that all Full Moons seem to appear larger when seen close to the horizon. This is an optical illusion because the observer is subconsciously comparing the Moon to terrestrial objects, in this case the CN Tower and skyline of Canada's biggest city.

ABOVE AND OPPOSITE NEAR LEFT:

Hale-Bopp Comet and Northern Lights, Alaska

Hale-Bopp was the brightest comet seen for many decades and probably the most widely observed comet of all time. It was discovered by amateur observers in July 1995 while it was still beyond the orbit of Jupiter.

OPPOSITE FAR LEFT:

Stars above East Twin peak, Alaska

East Twin is a rugged mountain in Chugach State Park. Rising to a height of 1790m (5872ft), it sees only a few dozen ascents by climbers each year. Long winter nights here give plenty of time for stargazing.

RIGHT:

Aurora Borealis, Fort Greely, Alaska

Fort Greely occupies a remote location in the interior of Alaska. The Northern Lights were regarded as sacred by Native Americans, who have numerous legends to explain them. The aurora was often believed to be human spirits in the sky.

PREVIOUS PAGE:
Milky Way over Crater Lake, Oregon, USA

Occupying a volcanic caldera formed 7700 years ago by a volcanic eruption, Crater Lake is a beautiful subalpine lake with exceptional water clarity. In spring, snow melts from the rim of the caldera and the complete arch of the Milky Way can be seen between opposite horizons. The lights of Crater Lake Lodge can be seen on the opposite rim.

OPPOSITE:
Total solar eclipse, Ritter, Oregon, USA

The total solar eclipse of 21 August 2017 crossed the entire width of the continental United States. This photograph of the eclipsed sun in the eastern sky was taken just after totality began at 10.20am. The Moon's shadow is racing eastwards across the Earth at some 3000 km/hour and we can see the sunlit land and sky in the distance, where the lunar shadow has not yet reached.

RIGHT:
Baily's Beads, total solar eclipse, Ritter, Oregon, USA

As a total solar eclipse ends, the Sun reappears shining first through the valleys between mountains on the lunar limb as beads of dazzling luminescence. These Baily's Beads are extremely bright and change very rapidly, lasting less than a second.

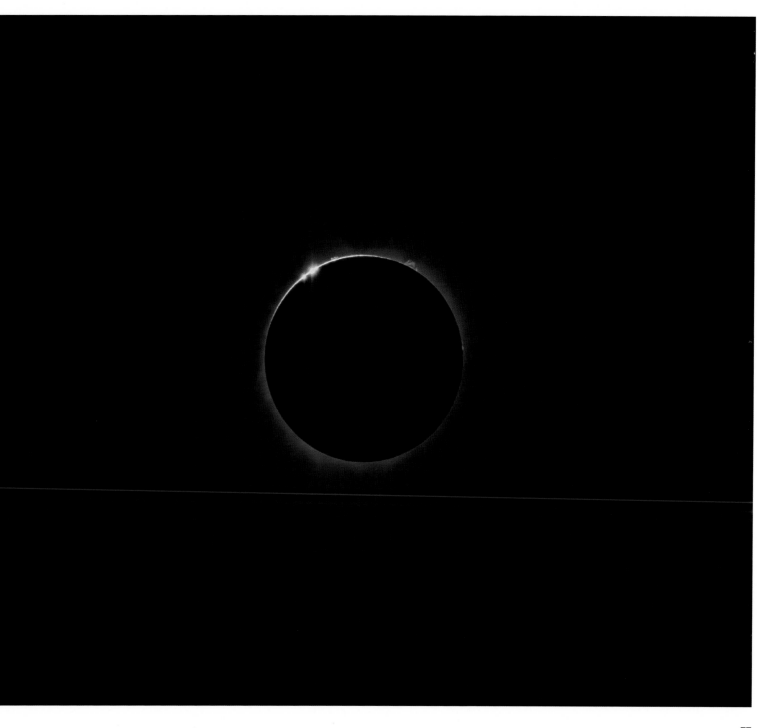

Total solar eclipse, Madras, Oregon, USA
Madras was a popular location for viewing the total solar eclipse on 21 August 2017. It is dangerous to view the partial phases of a solar eclipse without eye protection but once the sun is totally eclipsed, it is safe to view it with the naked eye.

ABOVE:
Hale Bopp Comet and Mauna Kea Observatories, Hawaii, USA
The Subaru Telescope, Keck 1 and Keck 2 telescopes and NASA Infrared Telescope Facility are located at an elevation of 4145m (13,600ft) near the summit of Mauna Kea, giving exceptionally clear observing conditions for most of the year.

ABOVE:
Badwater Basin, Death Valley, California, USA
Repeated cycles of flooding and evaporation, followed by
desiccation in the hot desert sun, have created these curious
salt flats. Death Valley is the largest National Park in the
contiguous United States, so there is no light pollution here.
In this image, the salt flats are illuminated by moonlight.

OPPOSITE:
Milky Way over Red Rock Canyon, California, USA
Located where the Sierra Nevada meets El Paso Mountains,
Red Rock Canyon features spectacular desert cliffs and buttes.
Although only two hours' drive from Los Angeles, the skies
of this State Park are dark enough to see complex structure
within the core of the Milky Way, seen here as it sets on an
early autumn evening.

Milky Way and Saturn,
Joshua Tree National Park, California, USA
Joshua Tree National Park is a vast protected wilderness in southern California with spectacular rock formations. A little cloud in the sky complements the Milky Way on this early morning at the end of April. In 2017, Saturn (the brightest object in the inset image) was close to the core of the Milky Way in Sagittarius.

Milky Way, Joshua Tree National Park, California, USA
Joshua trees are a species of tree-like yucca that is indigenous to the Mojave Desert. They make iconic foregrounds in this photograph of the Milky Way rising soon before dawn in spring. The core of the Milky Way lies in the southern constellation of Sagittarius.

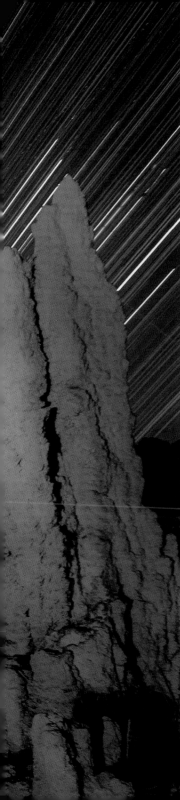

LEFT:

Star Trails, Mono Lake, California, USA

Mono is a soda lake fed by underground springs which are rich in dissolved minerals. Calcium carbonite is deposited around the springs to form tufa columns, which have been exposed to the air as the lake level has fallen. This image was taken looking north east across the South Tufa area, capturing the trails made by stars as Earth rotates on its axis.

ABOVE:

Moon ring, Arches National Park, Utah, USA

A moon ring or halo is an optical phenomenon formed as moonlight is refracted through countless hexagonal ice crystals suspended in Earth's atmosphere. The radius of the halo is approximately 22 degrees. Because red light is refracted slightly less than blue light, the inside of the halo appears reddish whilst the outside is tinged blue.

Orion, Sirius and Jupiter over June Lake, California, USA
Orion is directly above June Lake, the line of the three stars in its belt pointing to Sirius, the brightest star in the night sky, to its left. About three times brighter than Sirius is Jupiter, to the right of Orion in the constellation of Taurus.

Night Sky over Golden Gate Bridge, California, USA
San Francisco is the 13th most populous city in the United States but viewed from Fort Baker on the north side of the Golden Gate Bridge, the camera can record a good number of stars over the bright city lights.

Orion, Sirius and Jupiter over Lake Tahoe, California, USA

Straddling the borders of California and Nevada, Lake Tahoe is the largest alpine lake in North America. Orion occupies centre stage in this image. Jupiter, to Orion's right, is 588 million km (365 million miles) from Earth at its closest and its light takes 33 minutes to traverse that distance. To the left of Orion is Sirius, at a distance of 18 trillion km (11 trillion miles), whose light takes 8.6 years to reach us. Sirius is the 10th closest star to Earth (the closest being the Sun) but 138,000 times further away than Jupiter.

Milky Way over Canyonlands National Park, Utah, USA
Canyonlands is a vast wilderness of canyons, mesas, and buttes carved by the Colorado and Green Rivers into the Colorado Plateau of southern Utah. In spring, the core of the Milky Way lies low in the eastern sky before dawn, echoing the shape of Mesa Arch in this image. The centre of our galaxy, at a distance of 25,000 light years, is hidden from our view by huge clouds of gas and dust surrounding its core.

Geminid meteor shower over the Very Large Array Radio Telescope, Magdalena, New Mexico, USA

The Geminids, peaking annually on 13 December, are the brightest and most consistent meteor shower of the year. A total of 23 meteors are captured in this image, made at the peak of the shower in 2015. The meteors were photographed on a sequence of 334 30-second exposures, which were combined to show the meteor shower over a period of three hours.

ABOVE:

Annular eclipse over New Mexico, USA, 20 May 2012

An annular eclipse occurs when the Moon passes between the Earth and the Sun at a time when the Moon is further away than average in its orbit. The Moon appears smaller and does not completely cover the Sun so a bright ring of sunlight remains around it. Annular eclipses are slightly more common than total eclipses but less spectacular, as the Sun's corona and flares cannot be seen and the sky does not become dark.

RIGHT:

Crescent Moon and Venus over downtown St Louis, Missouri, USA

Gateway Arch in St Louis stands 192m (630ft) tall, making it the world's tallest arch. In this image the arch frames a two-day old Crescent Moon. The night side of the Moon can also be seen, faintly illuminated by sunlight reflected from Earth (earthshine). To the left is Venus, which if viewed through a telescope would show a similar crescent to that of the Moon.

OPPOSITE:

View from Clingman's Dome at night, Great Smoky Mountains National Park, Tennessee, USA

In the Appalachians of Tennessee and North Carolina, Great Smoky Mountains is the most visited National Park in the United States. Rising to 2025m (6644ft), Clingman's Dome is the park's highest mountain. In this five minute exposure looking north from the Dome's observation tower towards the densely populated lowlands outside the park, the brighter stars have formed short trails above bright city lights.

OPPOSITE:
Milky Way over Hunting Island, South Carolina, USA
The skeleton of a tree killed by seawater stands on the beach of Hunting Island, a lonely sentinel to the night sky. Looking south around midnight in summer, the core of the Milky Way is framed by broken cloud.

RIGHT:
Milky Way over New Smyrna Beach, Florida, USA
Shortly before sunrise, the core of our galaxy rises over New Smyrna Beach, a popular surfing destination just north of Cape Canaveral on Florida's east coast.

Milky Way over Cape Romano Dome House, Florida, USA

The Dome House was constructed in 1980 as a private holiday home but succumbed to coastal erosion and now stands in 2m (6ft) of seawater, 55m (180ft) from the shoreline of Cape Romano. The abandoned ruins make an evocative, otherworldly foreground to this image of the Milky Way setting into the Gulf of Mexico.

A panoramic view of the Milky Way Galaxy over Stage Harbor Lighthouse at Hardings Beach in Chatham, Massachusetts, USA
Stage Harbor Lighthouse was built in 1880, decommissioned in 1933 and is now a private residence. Looking out towards the Atlantic, there is no light pollution to interrupt the view as the core of the Milky Way rises in the eastern sky.

South America

The South American continent stretches far into the southern hemisphere, revealing aspects of the night sky that are never seen from Europe and North America.

When Ferdinand Magellan made his voyage of discovery around South America which culminated in the first circumnavigation of the Earth in 1522, he noted the presence of two 'clouds' in the southern sky. These now bear his name, the Large and Small Magellanic Clouds. Four centuries after his voyage, it was shown that they are independent galaxies outside the Milky Way.

Naturally, the Magellanic Clouds were known to Native Americans for thousands of years before European explorers first saw them. The greatest pre-conquest civilization in South America was the Incas, whose calendars were strongly tied to astronomy. Inca astronomers understood equinoxes, solstices and constructed elaborate monuments to observe and track these events.

Today, South America has some of the world's finest observing locations for professional astronomy. The Atacama Desert and adjacent Altiplano of northern Chile have almost no rainfall, low cloud cover and clear skies. High elevation means little atmospheric turbulence to interfere with observations. A major international collaboration has completed the European Southern Observatory's Very Large Telescope, soon to be surpassed by the Extremely Large Telescope. These huge instruments enable us to search for planets in other solar systems, study the structure of our galaxy in unprecedented detail and to look so far into space and see the universe as it was in its youth, some 13 billion years ago.

OPPOSITE:
Star Trails, Atacama, Chile
The Atacama is the driest desert in the world (outside the polar regions). Clear skies with very low humidity provide ideal observation conditions for the night sky. This long exposure captures the circles made by stars around the south celestial pole as the Earth rotates on its axis.

Llano del Hato National Astronomical Observatory, Merida, Venezuela

At an altitude of 3600m (11,800ft), Llano del Hato Observatory benefits from very low atmospheric turbulence. It is situated at a latitude of eight degrees north, making it the closest major astronomical observatory to the equator. Over the course of a year, Llano del Hato can observe over 90 per cent of the global night sky, more than any other facility in the world.

Ologa, Lake Maracaibo, Venezuela
Ologa is a fishing village on the shore of Lake Maracaibo,
a brackish inlet of the Caribbean Sea. Village houses are
built on stilts, which led to the name 'Venezuela' (meaning
'little Venice'). The particular geography of the lake creates
Catatumbo lightning on around 150 nights per year, making
this the most lightning-prone place in the world.

Stars over Tungurahua Volcano, Colombian Andes, Ecuador
Tungurahua means 'throat of fire', an apt name for this
active volcano. The current phase of eruptions began in 1999,
producing explosions, pyroclastic flows and lava flows. A
number of nearby towns and villages have been temporarily
evacuated. Rising to 5023m (16,480ft), Tungurahua used to be
capped by snow and a small glacier, which have now melted.

A starry night over Chimborazo, Ecuador
Located just one degree south of the Equator, the dormant volcano of Chimborazo is the highest mountain in Ecuador, with a summit elevation of 6263m (20,548ft). Thanks to Earth's equatorial bulge, Chimborazo is the world's highest mountain measured from the centre of the Earth (Mount Everest is higher measured from sea level). Climbing to the top of Chimborazo takes you as close to the stars as you can get without leaving Earth.

OPPOSITE:
Milky Way over Cordillera Huayhuash, Peru
The Huayhuash Trek is a demanding 130km (81-mile) hike between sheer mountain peaks and pristine glacial lakes, taking in elevations up to 5490m (18,011ft). There is little vegetation at these altitudes and no light pollution to interfere with the view of the Milky Way, high in the sky above the snow-capped Andes.

ABOVE:
Ishinca base camp, Cordillera Blanca, Peru
First quarter Moon illuminates base camp on the slopes of Ishinca in the Peruvian Andes. Rising to 5530m (18,143ft), the name Nevado Ishinca means 'snow-covered mountain'. Temperatures in the Ishinca Valley drop below freezing but lights within the climbers' tents give a warm glow.

NEXT PAGE:
Milky Way over Mount Auzangate, Cordilliera Vilcanota, Peru
At a latitude of 14 degrees south of the equator, the core of the Milky Way rises high in the sky. The core is not uniformly bright as parts of it are obscured by dark clouds of gas and dust. At the centre of the Milky Way lies a supermassive black hole, with an estimated mass four million times that of the Sun.

ABOVE:
Flyby of the International Space Station (ISS), Pedra Azul and Venda Nova do Imigrante, Espirito Santo State, Brazil
Rising in the west soon after sunset, the International Space Station make a bright trail of reflected sunlight across the sky, below the constellation of Orion. Launched in 1998, the space station has been continually inhabited since 2000, usually with a crew of six. At a cost of US$150 billion, the ISS has been described as the most expensive single object ever constructed.

OPPOSITE:
Dusk, Leblon Beach, Rio de Janeiro, Brazil
Here we can see the Leblon neighbourhood and beach with the Morro Dois Irmãos rock formation in the background. A magical time of the day when the artificial lights of the bustling city and the dim light of the night sky are balanced.

OPPOSITE FAR LEFT:

Milky Way core rising behind Pedra Azul, Espirito Santo, Brazil

Reaching a height of 1822m (5978ft), Pedra Azul ('Blue Stone') is a granite rock formation in the Atlantic forest ecological corridor of Brazil. The core of the Milky Way rises in the early hours of the morning, with Jupiter shining brightly against it.

OPPOSITE NEAR LEFT:

Milky Way over Jalapao State Park, Brazil

Deep in the interior of Brazil, Jalapeno State Park is an area of cerrado (grassland), dunes, gallery forest and wetland. Far from any towns, the sky is dark enough here for detail to be visible even in the outlying parts of the Milky Way.

RIGHT:

Milky Way over Golden Tree, Cerrado Biome, Brazil

The golden trumpet tree is native to the cerrado of Brazil, a dry grassland habitat found in the southern province of Mato Grosso do Sul. The photographer has painted this tree with light from a torch to show its spectacular foliage against the Milky Way.

ABOVE AND LEFT:

Milky Way and Magellanic Clouds over Uyuni salt flats, Bolivia

Uyuni salt flats are so exceptionally smooth that rain transforms them into a vast mirror. Above the salt flats is the Milky Way and two independent galaxies – the Large (inset) and Small Magellanic Clouds.

OPPOSITE LEFT:

Zodiacal light over Uyuni Salt Farm, Bolivia

Our solar system contains interplanetary dust. After sunset and before sunrise, a pale band of light, called 'zodiacal light', can sometimes be seen extending up from the horizon. This is caused by sunlight scattered from this dust.

OPPOSITE RIGHT:

Lake and Cuernos Mountains, Torres del Paine, Chile

The faint band of the Orion Arm of the Milky Way rises above the Cuernos Mountains, with the bright star Sirius to its left.

The Milky Way over the Atacama desert, Chile
The complete arch of the Milky Way frames a remote desert road near San Pedro de Atacama in Chile's Antofagasta Region. At an elevation of 2100m (6890ft) and with just 38mm (1.5in) of rain a year, there is little cloud cover or atmospheric turbulence to obscure the night sky.

Orion, Sirius and Pleiades above ice penitentes in the Andes, Chile
Penitentes are tall thin blades of hardened snow or ice, closely spaced with the blades oriented towards the general direction of the Sun. The ice form in a dry climate where the local dew point remains below freezing, causing ice to be lost by sublimation (evaporation without melting). Looking north, Orion, Sirius and the Pleiades are in the sky above.

LEFT:

Centaurus above the European Southern Observatory's telescope at La Silla Observatory, Chile
Surrounded by the mass of stars making up the Milky Way, the two bright stars directly above the telescope dome are Alpha and Beta Centauri.

OPPOSITE LEFT:

European Southern Observatory, Chile
The European Southern Observatory is a 16-nation collaborative research project to study the southern hemisphere skies. It is located in the Atacama Desert of northern Chile to take advantage of the clear skies and low atmospheric turbulence. Currently under construction is the Extremely Large Telescope with a light-gathering mirror 39m (128ft) in diameter, scheduled for completion in 2024.

OPPOSITE RIGHT:

Star trails above the Very Large Telescope at Paranal Observatory, Chile
The European Southern Observatory's Very Large Telescope comprises four individual telescopes, each with a primary mirror 8.2m (27ft) in diameter. Located at an altitude of 2635m (8645ft) in northern Chile, the facility benefits from a very dry climate with very few clouds, enabling 340 observing nights a year.

PREVIOUS PAGE:

Star trails over ALMA radio astronomy antennas, Chile
The Atacama Large Millimeter Array (ALMA) is an international research facility located at an elevation of 5000m (16,000ft) on the arid Chajnantor plateau in the Chilean Andes. The high elevation and low humidity enable research into the birth of stars during the early stages of the universe.

ABOVE:

Meteor and crescent Moon over Pan de Azucar National Park, Chile
Meteors originate as grains of interplanetary dust, usually associated with comets or asteroids, which strike Earth's atmosphere at typical speeds of 20km/ second. The brief trail of light that we see is caused by friction with air molecules as the particle is vaporized.

OPPOSITE:

Moai at moonrise, Easter Island, Chile
Moai are monolithic human figures carved by the Rapa Nui people on Easter Island between the 13th and 15th centuries. Hundreds were transported from where they were carved and set on stone platforms around the island. The Moon rises over these highly stylized human figures, with overly large heads.

Moonrise over Lauca National Park, Chile
Located in the far north of Chile, Lauca National Park is an arid area of altiplano and mountains. Cerro Cosapilla is one of many huge volcanoes in this part of the Andes.

Milky Way over Cerro Armazones, Chile
The arch of the Milky Way, with the complex structure of stars and nebulae in its core, above these towers on Cerro Armazones in the Chilean desert. This is the location of the European Southern Observatory's Very Large Telescope, which is ideally placed to study this complex area of our galaxy.

ABOVE:

Aurora Australis over Andes, Chile
Chilean Patagonia extends to a latitude of more than 53 degrees South, from where the Aurora Australis (Southern Lights) can sometimes be seen over the snow-capped Andes.

RIGHT:

Moon over Punta del Diablo, Uruguay
The last hint of sunlight is draining from the sky over Punta del Diablo beach, revealing the stars and Moon. A little artificial light illuminates the texture of the foreground dune.

OPPOSITE:

Moonrise over the Beagle Channel, Ushuaia, Argentina
In this picture, taken soon after sunset, we see the sky tinged pink around the Full Moon as a result of sunlight from below the horizon scattered in Earth's atmosphere, a phenomenon known as the 'Belt of Venus'.

Moon and Venus over Buenos Aires, Argentina
Just before sunrise in December, the waning crescent Moon and the planet Venus lie close in the dawn sky over the Rio de la Plata estuary. The night side of the Moon is lit by Earthshine, which is tricky to photograph in combination with the sunlit crescent as it is about 10,000 times dimmer than a Full Moon.

LEFT:

Fitz Roy Massif and Glacial Lake, Patagonian Cordillera, Chile

Mount Fitzroy lies on the border of Chile and Argentina in the southern Patagonian icefield. It was named in 1877 but not climbed until 1952. In this picture it is brightly illuminated by gibbous moonlight under the dazzling southern hemisphere night sky.

RIGHT:

Comet McNaught over Patagonia, Argentina

Comet McNaught was discovered in August 2006 and went on to become the brightest comet observed in the last 50 years. At its peak in January 2007, it was brighter than any of the planets, had a tail stretching 35 degrees across the sky and was visible in broad daylight. The path of this comet around the Sun meant that observers in the southern hemisphere had the best view.

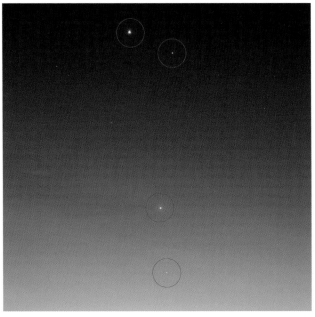

ABOVE AND OPPOSITE LEFT:

Four planets over the Atlantic, Buenos Aires, Argentina
It is unusual to see four planets visible to the naked eye close together in the sky. On 5 May 2011, just before sunrise, Venus, Mercury, Jupiter and Mars (from top to bottom in this image) were in conjunction. Venus and Jupiter are much the brightest, whilst Mars is almost lost in the pre-dawn pink glow.

LEFT:

Lunar halo, Punta Piedras, Argentina
A 22 degree halo (also known as a 'winter halo') around the Moon is caused by refraction of moonlight within ice crystals in Earth's atmosphere. The inside of the halo is tinged red as the angle of refraction for red light is 21.54 degrees, compared to 22.37 degrees for blue light.

OPPOSITE RIGHT:

Milky Way over Iguazu Falls, Argentina
On the border of Argentina and Brazil, the Iguazu Falls are the largest and arguably the most beautiful waterfall system in the world. Higher than Niagra and with a greater volume than Victoria, the falls are situated in lush subtropical rain forest with little light pollution to obscure the Milky Way.

ABOVE AND LEFT:

Scorpius over Mercedes, Buenos Aires Province, Argentina
The zodiacal constellation of Scorpius is one of the most prominent in southern hemisphere skies. Its brightest star is Antares, circled in the above image, which is a binary star (a star system consisting of two stars that orbit around their common centre of mass). The larger star of the pair is a red supergiant, 10,000 times more luminous than our Sun, which is destined to end its life as a supernova.

RIGHT:

Milky Way and southern twilight, Pebble Island, Falkland Islands
Located in the southern Atlantic Ocean, The Falkland Islands have only 3000 human inhabitants and hence very dark skies. This image was captured around midnight in late November, when twilight glow in the southern sky lasts throughout the night. Looking west, the Milky Way stretches across the image and a satellite trail is visible at top left.

Australasia

Australia is sometimes described as the world's most urbanized major country, with cities accounting for 90 per cent of the population. The Outback, however, represents 70 per cent of its land area and has dark skies to equal any on Earth. Striking natural features such as Uluru in Northern Territory and The Pinnacles in Western Australia are ideal for sky watching.

One of the most southerly countries in the world, New Zealand's glaciated mountains, stunning coastline and small population make it ideal for night sky photography. It is probably the best place for the Aurora Australis, or Southern Lights, which are less commonly observed than their northern counterparts but equally spectacular.

Aboriginal Australians have a rich mythology for the heavens. The core of the Milky Way is seen as an emu in the sky; its outline defined by dark nebulae seen against the bright background of stars in our galaxy. Annual events such as gathering malleefowl eggs were scheduled around the rising or setting of corresponding constellations in the sky. The Maori of New Zealand, in common with other Polynesians, were great navigators and undoubtedly used the stars to guide their voyages across the Pacific.

Today, the 3.9m (12.8ft) equatorially mounted telescope operated by the Australian Astronomical Observatory is the largest observation facility in Australasia. In recent decades, bright comets have favoured observers in the southern hemisphere and Australian astronomer, Robert McNaught, has discovered 82 comets and 483 asteroids. In 2007, one of the comets bearing his name became the brightest seen in the last half century.

OPPOSITE:

The Milky Way framed by a Cave, Wellington, New Zealand

The core of the Milky Way lies low in the western sky during September as seen from New Zealand, enabling the photographer to capture it neatly framed by the entrance to this coastal cave. A little cloud movement has occurred during the long exposure needed to collect light from the Milky Way.

Partial solar eclipse through bushfire smoke over Broken Bay, Sydney, Australia
The total solar eclipse of 4 December 2002 was seen in several southern African countries, after which it crossed the Indian Ocean to reach the coast of South Australia at sunset. A partial eclipse was visible from Sydney late in the evening, where a thick plume of bushfire smoke has attenuated the sun's glare sufficiently for the eclipse to be photographed without a filter.

LEFT:

Venus and Jupiter in conjunction over Uluru, Australia

Sacred to the Anangu Aboriginal people, Uluru is the largest monolith in the world. Venus and Jupiter can be seen close together in the sky about once each year, outshining all the stars. Although Jupiter is by far the largest planet, Venus is 15 times closer to Earth and appears five times brighter.

BELOW:

Milky Way over Shipwreck Coast, Victoria, Australia

The Milky Way core lies low in the sky over the impressive rock stacks of Victoria's Shipwreck Coast between Cape Otway and Port Fairy. Bands of dust and glowing nebulae are seen along the band of billions of stars that is the view of our own galaxy from the inside.

The Pinnacles at night, Nambung National Park, Western Australia
The Pinnacles are thousands of weathered limestone formations within Nambung National Park in Western Australia. The geological mechanism by which they formed is not fully understood. The tallest pinnacles reach a height of up to 3.5m (11.5ft) above the yellow sand base and make an impressive foreground to the dazzling southern hemisphere sky.

OPPOSITE LEFT:

Milky Way over Uluru, Australia

Situated in the geographical heart of the continent, Uluru is Australia's best known, and most instantly recognizable, natural feature. Situated 450km (280 miles) from the nearest large town of Alice Springs, there is very little light pollution here to interfere with this magnificent view of the heart of our galaxy.

OPPOSITE RIGHT:

Comet Lovejoy over the Southern Ocean, Cape Schanck, Victoria, Australia

On 16 December 2011, Comet Lovejoy passed through the Sun's corona which was expected to destroy it. However, the comet survived and was photographed a week later just before sunrise, its long tail stretching away from the Sun.

ABOVE:

Comet McNaught over the outback, Mildura, Victoria, Australia

'The Great Comet of 2007' is the best known of over 50 comets discovered by Australian astronomer Robert McNaught, as it became the most brightest comet seen since 1965. Comets are made mostly of ice which vaporizes as they approach the Sun.

LEFT:
Night Sky above Wellington, New Zealand
The bright city lights of Wellington are shrouded in fog on a chilly May evening, allowing the stars and faint band of the Milky Way to be seen from Wrights Hill overlooking the city.

NEXT PAGE:
Milky Way and Magellanic Clouds over Wainuiomata coast, New Zealand
In this remote location near Baring Head, the core of the Milky Way appears very bright as it rises in the east, whilst the Large and Small Magellanic Clouds take centre stage in this image. These are independent galaxies outside the Milky Way, at distances of 158,000 and 199,000 light years respectively.

OPPOSITE LEFT:

**Milky Way and Jupiter,
Whangapoua Beach,
New Zealand**

Great Barrier Island is
a designated Dark Sky
Sanctuary, with a population
of 1000, no mains electricity
and almost no light pollution
to obscure the night sky. The
ribbon of light reflected by
the stream echoes the shining
path of the Milky Way, with
Jupiter shining brightly near
the centre of the frame.

OPPOSITE RIGHT:

**Total solar eclipse,
Northern Cook Islands**

The total solar eclipse of 21
July 2009 was the longest of
the 21st century, with up to
6 minutes and 39 seconds of
totality along its path. After
crossing China, the eclipse
path traversed the western
Pacific to reach the Northern
Cook Islands before sunset.
Looking west from on board
a ship, we see the Moon's
shadow leaving the horizon as
totality is about to end.

RIGHT:

**Moonset, Brooklyn Wind
Turbine, Wellington,
New Zealand**

Prominent in this view is the
bright crater Tycho on the
Moon's southern hemisphere
(towards top left on the
Moon's disk). It was formed
by a huge meteorite impact
about 108 million years ago.
Rays of ejected material
can be seen radiating
from the crater.

Milky way over Castle Rock, Castlepoint, Wairarapa, New Zealand
Castle Rock towers 162m (532ft) above the beach of Castlepoint in Wairarapa, New Zealand. From this viewpoint looking east out to sea, the complex structure of the Milky Way is well shown, with dark nebulae obscuring our view of the stars in parts of its core.

Moonrise over Lake Tasman, Aoraki/Mount Cook National Park, New Zealand
Orion is high in the sky on a summer evening as the moon rises over the snow-capped mountains in the Tasman Valley. New Zealand's longest glacier, the Tasman glacier, discharges into this lake, calving scattered icebergs which can be seen floating away down the lake.

ABOVE AND RIGHT
Milky Way and Canopus, Cape Palliser, New Zealand
The lighthouse at Cape Palliser forms a guiding light to the Milky Way arching across this 180 degree panoramic image. On the left are the Large and Small Magellanic Clouds along with Canopus, the second brightest star in the night sky. Canopus (seen in the above image) is an aged star, 310 light years away and radiating 10,000 times the luminosity of our Sun.

OPPOSITE:

Aurora Australis, Wellington, New Zealand
Red aurorae are unusual, being formed by excited atomic oxygen atoms at the highest altitudes of our atmosphere, under conditions of intense solar activity. This image was captured through a break in the clouds on 17 March 2015; a night of intense auroral activity in the southern hemisphere.

ABOVE:

Milky Way, Cable Bay, New Zealand
Located near Nelson on New Zealand's South Island, Cable Bay forms a perfectly balanced foreground to this view of the Milky Way. The galactic core is high in the night sky, with its light reflecting in the bay below just before moonrise.

NEXT PAGE:

Aurora Australis, Queenstown, New Zealand
The yellow and pink colours of this spectacular aurora are the result of mixing of the primary auroral colours of red, green and blue during an intense solar storm. Looking due south, the Large and Small Magellanic Clouds are visible above the shimmering curtain of light.

Milky Way, Tasman Valley, Aoraki/Mount Cook National Park, New Zealand

In this magnificent image, the Milky Way frames the two Magellanic Clouds, below which the camera has recorded a faint aurora in the southern sky. The Magellanic Clouds are irregular dwarf galaxies, of which the smaller one appears to be in orbit around the larger.

ABOVE:

Milky Way, Church of the Good Shepherd, Lake Tekapo, New Zealand

Lake Tekapo is situated in the centre of South Island, surrounded by spectacular natural beauty. Built in 1935, the Church of the Good Shepherd is one of the most photographed In New Zealand. On a clear night, the stars and the Milky Way shine brightly above.

Asia, Middle East, Far East

Astronomy has been practised in India, China and the Middle East for thousands of years. Indian astronomy has its roots in prehistory and flowered when Europe was in its Dark Ages. At the beginning of the eighteenth century, the Maharaja of Jaipur constructed five huge observatories containing complex instruments which could be aligned with various celestial bodies. Cosmological ideas developed in India went on to influence the development of astronomy throughout Asia and Europe.

Astronomy formed the basis of both Chinese and Islamic calendars, each of which adopt the lunar month as their fundamental unit. Chinese and Arabian scholars recorded astronomical events that seem to have gone unnoticed by observers in Europe, such as a supernova in 1054.

Ancient Chinese astronomers were the most accomplished of the ancient observers of the skies. They understood the nature of eclipses and could predict them, recording hundreds of observations of solar and lunar eclipses starting in 750 BC.

Arabic astronomers were likewise meticulous in their recording. Today, we continue to use the names given by Arab astronomers to many of the stars in the sky, such as Aldebaran (the brightest star in Taurus), Altair (the brightest star in Aquila) and Deneb (the brightest star in Cygnus).

More than half the world's human population lives in Asia, increasingly in cities. However, Asia also has huge deserts and the world's greatest mountain range, where city lights are far away and the wonders of the night sky can be truly appreciated.

OPPOSITE:
Ursa Major rising above Himalayan peaks and the village of Gorak Shep, Sagarmatha National Park, Nepal
At an elevation of 5150m (16,896ft) above sea level and surrounded by the world's highest mountains, Gorak Shep is the final stop for climbers before reaching Everest Base Camp. The familiar shape of Ursa Major (also known as 'The Plough' or 'Big Dipper') is high in the sky and Polaris, the North Pole star, is just above the steep peak on the left. Arcturus, the brightest star in the northern celestial hemisphere, can be seen on the right.

Night sky over Gunung Bromo, Java, Indonesia
Mount Bromo is an active volcano in Java's Bromo Tengger Semeru National Park. Indonesia has around 130 active volcanoes, more than any other country in the world, located around the Pacific Ring of Fire. When not erupting, Mount Bromo is a popular tourist destination.

Orion from Mount Kinabalu, Sabah, Malaysia
A huge granite massif rising to 4095m (13,435ft), Mount Kinabalu in Borneo is South-East Asia's highest peak. This image is taken from above Laban Rata Guest House, located 3272m (10,735ft) above sea level. Orion takes centre stage with its bright star Betelgeuse distinctly red in colour (circled in the above image). Sirius is low down near the horizon.

Stars over Tioman Island, Malaysia
Tioman Island is located 32km (20 miles) off the east coast of Malaysia. The island is densely forested and sparsely populated, so apart from beach chalets there is little artificial light to obscure the night sky.

The Milky Way above Lam Isu Reservoir, Kanchanaburi, Thailand

Trees frame the core of the Milky Way above Lam Isu Reservoir. There are two planets in this Image: Jupiter, seen against the Milky Way (see right in the inset image) and Saturn, above and just to the right of the mountain peak (see left in inset image). The distinctive pyramidal mountain on the shore of the reservoir is a popular location for photographing the Milky Way.

OPPOSITE:

Milky Way over Koh Hai Island, Thailand

Koh Hai is an idyllic island of white sand beaches. This picture shows the familiar form of the Milky Way core, its bright band of stars interrupted by dark clouds of gas and dust.

RIGHT:

Orion, Sirius and Canopus over the Bay of Bengal from Ngapali, Myanmar

The constellation of Orion lies near the centre of this image, not far from setting in the west. Sirius, the brightest star, is to its left and slightly above. Low down on the far left is Canopus, the second brightest star in the night sky. Canopus lies deep in the southern hemisphere sky and can never be seen from north of latitude 37 degrees North. Here in the tropics it makes a striking pair with Sirius.

NEXT PAGE:

Geminid meteor shower, Yunnan province, China

The meteors in this image were captured on a series of frames taken over three hours before dawn on 14 December 2012. Meteor showers are often most intense in the early hours of the morning as the pre-dawn side of the Earth is forward-facing in its orbit, travelling at 30km (20 miles)/second, so more meteor collisions occur than before midnight.

LEFT:

Orion and Sirius rise above the Yunnan Astronomical Observatory, China

This observatory is part of the International Burst Observer and Optical Transient Exploring System, studying optical emissions from gamma ray bursts that occur in the universe. The most noticeable object in the sky is the Orion Nebula, which is the closest region of massive star formation to Earth at a distance of 1,344 light years.

RIGHT:

Crescent Moon, Venus and Saturn, Patan, Nepal

The planet Venus can appear in the sky as a 'morning star' or an 'evening star', depending on its orbital position relative to Earth. In this case, Venus is seen after sunset, close to its eastern elongation from the Sun. Almost halfway between Venus and the waxing Crescent Moon is Saturn, viewed across the width of the solar system and much fainter.

FAR RIGHT:

Orion rises above Himalayas, Sagarmatha National Park, Nepal

Clear mountain air is ideal for star-gazing as atmospheric turbulence is reduced at high elevation. The mountain is illuminated by moonlight as Orion rises beyond. The bright blue star directly above the mountain peak is Rigel, Orion's brightest star.

LEFT:

Conjunction of Venus and Jupiter, Lake Namtso, Tibet, China

At an elevation of 4720m (15,486ft) Lake Namtso is one of the three holy lakes in Tibet, with crystal-clear blue water. Venus (on the left of the pair in conjunction) is our nearest planetary neighbour and is permanently covered in highly reflective white cloud, which accounts for its dazzling brilliance in this image.

ABOVE:

Milky Way over Namgyal Tsemo Gompa Buddhist Monastery in Leh, Ladakh, India

This image shows great detail in the core of the Milky Way, acquired during a long exposure with the camera on a tracking mount. There are estimated to be some 250 billion stars in our galaxy, all orbiting around a supermassive black hole hidden from view by the stars, gas and dust in this image.

181

LEFT:

Milky Way over Leh city in Leh Ladakh, India

Leh is a high desert city in the Himalayas of northern India. Originally a stop for trading caravans, Leh is now known for its Buddhist sites and nearby trekking areas.

RIGHT:

Milky Way over Chandratal Lake, Himachal Pradesh, India

At an elevation of 4300m (14,108ft) in India's Himalayas, Chandratal is known as the 'Lake of the Moon' from its crescent shape. The peaks in this image are illuminated by an almost Full Moon, so only a faint band of the Milky Way is visible in the sky beyond.

ABOVE AND OPPOSITE LEFT:
Sirius in Canis Major and Orion above Alborz Mountains, Iran
Sirius is referred to as the 'dog star' as it is by far the most prominent star in the constellation of Canis Major (the Great Dog, which follows Orion, the Hunter, into the night sky).

LEFT:
Milky Way over Garmeh Oasis, Dasht-e Kavir, Iran
Dasht-e kavir is a large desert in the middle of the Iranian plateau. The Milky Way stands out in a clear sky studded with stars, with just a little distant light pollution on the horizon.

OPPOSITE RIGHT:
Leaked rocket fuel over Alborz Mountains, Iran
The white patch in this image is rocket fuel from an Atlas V rocket reflecting sunlight in the night sky. The other bright object in the image is the planet Venus.

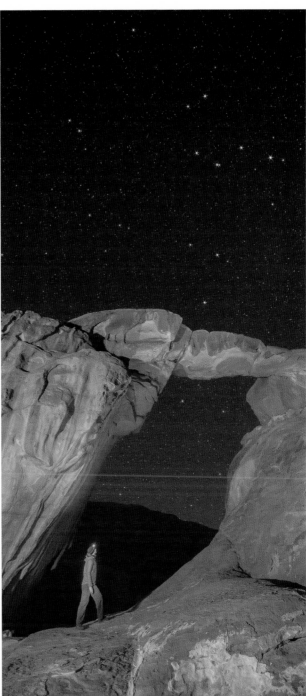

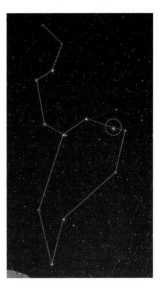

ABOVE AND LEFT:

Observing Leo, Um Frouth rock arch, Wadi Rum, Jordan

The prominent constellation of Leo has just risen above the arch. Leo is one of the most distinctive constellations, its shape reminiscent of a crouching lion. Its brightest star is Regulus.

FAR LEFT:

Milky Way, Venus, Jupiter and Saturn over Wadi Rum, Jordan

Taken just before dawn, this image shows the arch of the Milky Way from the galactic core on the right to the fainter outlying parts of our galaxy on the left. Jupiter can be seen against the Milky Way, with fainter Saturn just below it. Venus is brighter than both, having just risen as pre-dawn glow starts to lighten the eastern sky.

187

Africa

Recorded African astronomy has its most ancient roots in Egypt. The pyramids at Giza, built in the third millennium BC, are precisely orientated towards the pole star as it was 4500 years ago. Similarly, the Great Temple at Karnak was aligned on the rising of the midwinter Sun at the time it was built. The annual first rising of stars and constellations was of great importance to the Egyptians, for example the first rising of Sirius in the dawn sky presaged the flooding of the Nile. In the second century AD, the Egyptian astronomer Ptolemy developed an Earth-centred model of the Universe that became accepted in Europe for 1200 years.

Today, the pyramids are just a short distance from Cairo, a modern city of almost 20 million people, so the night sky is hardly visible at these ancient monuments. Most of Africa, however, remains a rural society with little light pollution. The majority of the continent is desert, semi-desert and savanna, which have low cloud cover.

In sub-Saharan Africa, the distinctive shapes of drought-resistant trees such as baobabs and quiver trees make compelling foregrounds to the night sky. Much of southern Africa is ideal for appreciating the full glory of the Milky Way. The core of our galaxy lies in the southern celestial hemisphere. Compared to the outlying arms of our galaxy that are more readily seen from the Northern Hemisphere, the core is brighter and has much more complex visible structure. To stand under African skies after dark with the Milky Way overhead is a reminder of our place in the universe.

OPPOSITE:
Milky Way over the Avenue of baobabs, Madagascar
Near Madagascar's west coast, huge Grandidier's baobabs tower over all other trees, silhouetted against the star-studded southern hemisphere night sky and the pale band of the Milky Way. Living in Africa, our earliest human ancestors would have been much more familiar with views like this than most of us today.

Night sky over Erg Chebbi, Morocco
Erg Chebbi is a semi-arid area of impressive wind-blown sand dunes on the edge of the Sahara Desert. Moonlight illuminates this lonely tree against a backdrop of golden dunes.

ABOVE AND LEFT:

**Stars of constellation Centaurus,
Tassili National Park, Algeria**
Centaurus is a southern hemisphere constellation, just visible in
its entirety from North Africa when it reaches its highest point
above the horizon in May. Barely above the line of sandstone
cliffs in this image and circled is the constellation's brightest star
Alpha Centauri, which is the closest star system to our Solar
System, at a distance of 4.3 light years (40 million million km).

FAR LEFT:

Milky Way over Erg Chebbi, Morocco
Viewed from high on a sand dune the core of the Milky Way
rises in the early hours of a late winter morning. The nearest
towns are Erfoud and Rissani, some 50km (32 miles) away,
whose lights are illuminating the sky on the right of this image.

LEFT:
Pyramids at night, Giza, Egypt
Astronomy was of great importance to the builders of the pyramids. The ancient Egyptians appear to have constructed their 365-day civil calendar with its New Year corresponding with the date of Sirius's return to the night sky in July. This astronomical event also marked the flooding of the Nile in Ancient Egypt. It is thought that astronomical alignments may have played a role in the orientation of the pyramids themselves.

NEXT PAGE:
Milky Way and Large Magellanic Cloud over Mount Kilimanjaro, viewed from Amboseli, Kenya
Towering over the surrounding savanna, Kilimanjaro is Africa's tallest mountain. The Milky Way can be seen to its left, with the Carina Nebula visible within it. This is a large, complex area of bright and dark nebulosity including areas of star formation and young star clusters.

LEFT:

Masai Tribesman under Milky Way, Kenya
A Masai tribesman in traditional costume stands outside his hut at night near the Masai Mara. Masai are pastoralists who practise a semi-nomadic lifestyle on the East African savanna. The core of the Milky Way is overhead and a meteor to its left.

BELOW:

Milky Way over Acacia Tree, Masai Mara, Kenya
Acacia are one of the most distinctive trees of the African savanna. In game reserves such as the Masai Mara, there is little light pollution to interfere with the brilliance of the Milky Way.

Large Magellanic Cloud over Masai Mara, Kenya
Giraffes and wildebeests must remain constantly alert at night to guard against predation. The Large Magellanic Cloud visible in the sky above them is a barred spiral galaxy with about one per cent of the mass of our own Milky Way. Located close to the south celestial pole, it can only be seen from latitudes south of the Tropic of Cancer.

OPPOSITE:

Night sky over Mount Kilimanjaro, viewed from Amboseli National Park, Kenya
Mount Kilimanjaro in Tanzania is considered to be the world's highest free-standing mountain, in that it is not part of a mountain range. The summit of this dormant volcano is glaciated but about 80 per cent of the ice is estimated to have been lost in the last 100 years.

Quiver Trees and three galaxies, Keetmanshoop, Namibia

The stark outlines of dolerite rocks and quiver trees are silhouetted against the exceptionally dark night sky in southern Namibia. In this view of the southern celestial hemisphere, the core of the Milky Way is on the right, with the Large and Small Magellanic Clouds on the left. All the foreground stars – of which many more have been recorded by the camera than are visible to the unaided eye – are in our own galaxy.

RIGHT:

Milky Way over Dead Vlei, Soussvlei, Namibia

Dead Vlei is a pan in the Namib Desert which was formerly irrigated by occasional flood waters, allowing trees to grow there. Around 900 years ago, migrating sand dunes cut off the water supply and the trees died. Almost a millennium later, their skeletons stand desiccated in the dry desert air under the sparkling southern hemisphere sky.

FAR RIGHT:

Star trails, Dead Vlei, Sossusvlei, Namibia

This image was captured by leaving the camera to photograph the stars for four hours in the desert bathed in the light of a four-day old Crescent Moon. The south celestial pole is directly behind the dead tree, resulting in concentric circles of star trails radiating outwards as Earth spins on its axis.

OPPOSITE:

Baobab trees and Milky Way, Makgadikgadi National Park, Botswana

Baobabs are succulent trees; their huge swollen trunks enable them to store water through the long, dry season. Having shed their leaves, they make strong silhouettes against the dark skies of this remote part of north-eastern Botswana.

RIGHT:

Milky Way over the Southern African Large Telescope, Northern Cape Province, South Africa

The Southern African Large Telescope is a 10-m (33-ft) class optical telescope designed mainly for spectroscopy. Located in the semi-arid Karoo region, there are few clouds and little light pollution. In this image, the Large Magellanic Cloud is almost over the observatory, the Small Magellanic Cloud to its left and the Milky Way above.

Comet McNaught, Cape Town South Africa
In January 2007, Comet McNaught reached a peak brightness greater than any of the planets and could even be seen in broad daylight. It is seen here at sunset, just six days after its closest approach to Earth. The comet then left the inner Solar System and will not return for 93,000 years.

Night sky over Blyde River Canyon, Mpumalanga, South Africa
Rising to 3482m (11,424ft), the Drakensberg Mountains are the eastern portion of the Great Escarpment, which encloses the central Southern African plateau. Clouds low in the sky are lit by distant light pollution, whilst the Milky Way and countless stars are visible above the clouds.

Appendices

Solar eclipse on 1 August 2008 from Russia
The total eclipse of the sun of 1 August 2008 began at sunrise in northern Canada, crossed Greenland and then Siberia before ending in China at sunset. In this image from Russia, about one-quarter of the Sun is hidden behind the Moon.

Solar eclipse before totality
About 80 minutes after the eclipse began, only a crescent of sunlight remains. The quality of light changes very rapidly in the last few minutes before totality, with shadows taking on a strange quality and the air turning cold.

Solar eclipse final stages
As the last 0.01 per cent of the Sun disappears from view, a 'diamond ring' effect is created around the Moon. The ring is formed by the Sun's corona, which comes into view when the much brighter solar disk is completely eclipsed.

Solar eclipse sun reappearing
The total eclipse lasted just two minutes, after which the Sun reappeared from behind the lunar limb as a bead of dazzling brightness. In this image we can see structure in the Sun's corona, warped by the intense solar magnetic field.

A Lunar Month

The Moon takes 27.5 days to orbit Earth but thanks to Earth's orbit around the Sun, it takes 29.5 days to complete its phases. Whatever the phase of the Moon, it displays the same familiar face to Earth, having become tidally locked to our planet early in its history.

Star Charts

There are many ways to divide up the night sky. These charts use one of the simplest, splitting it into six segments of roughly equal extent: four along the celestial equator, and one each for the poles. The chart on this page shows the night sky around the north celestial pole

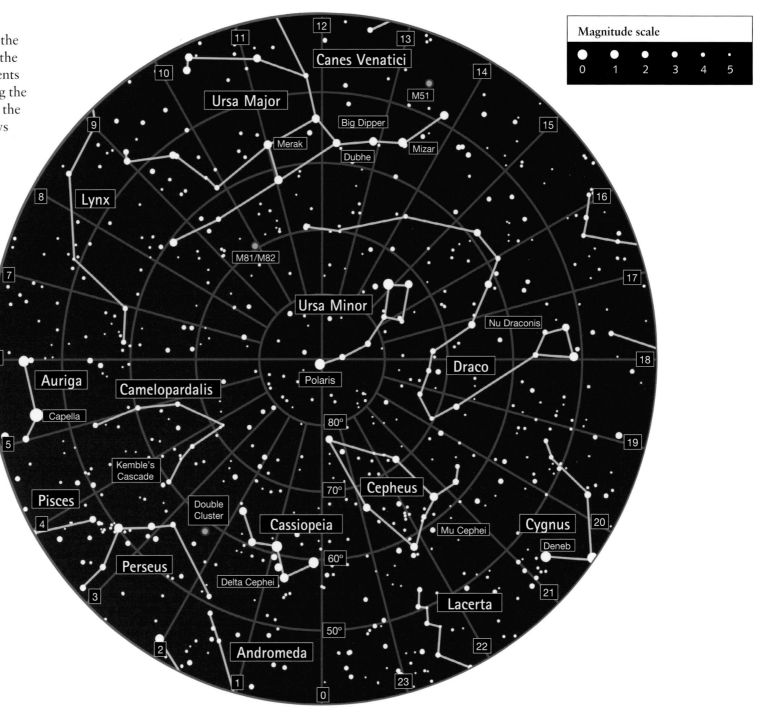

Magnitude scale

0 1 2 3 4 5

The Northern Polar Night Sky

The stars in this chart are always visible throughout the night to observers in northern latitudes such as Britain. The north celestial pole is marked by a second magnitude star, Polaris, which until the advent of modern Global Positioning Systems, was extremely useful to navigators.

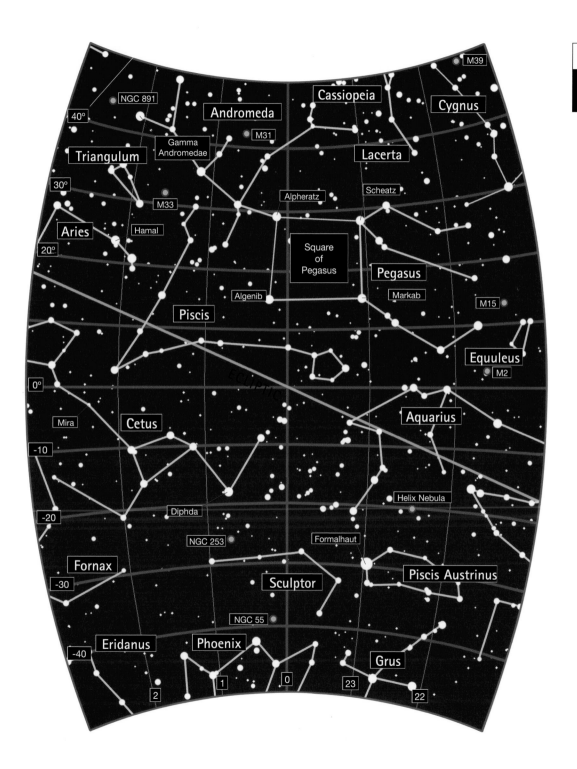

Magnitude scale

0 1 2 3 4 5

The Night Sky around the Square of Pegasus
Made up of four almost equally spaced stars, the Square of Pegasus is the most obvious feature in this part of the sky. To its left is the constellation of Andromeda, best known for its galaxy M31, which is bigger than the Milky Way and the most distant object visible to the naked eye.

Magnitude scale

0 1 2 3 4 5

Perseus

Capella

Algol

Auriga

M38

Pleiades
star cluster

Castor

Pollux

Gemini

M35

Taurus

M44

Hyades
star cluster

Cancer

Aldebaran

Orion

Canis Minor

Betelgeuse

Procyon

Hydra

M78

Orion's Belt

Monoceros

Orion
Nebula

Rigel

Sirius

Lepus

M41

M79

Eridanus

Canis Major

Puppis

Columba

Caelum

Horologium

**The Night Sky
around Orion**
Orion dominates the
winter sky in the northern
hemisphere and the southern
hemisphere summer
sky. To its north are the
easily recognized zodiac
constellations of Gemini
and Taurus, whilst to the
south is Sirius, the brightest
star in the night sky.

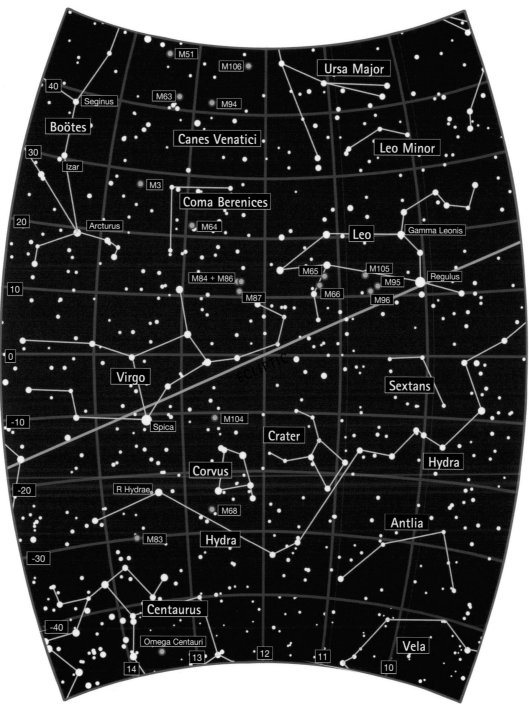

Magnitude scale

0 1 2 3 4 5

M51
M106
M63
M94
40
Seginus
Boötes
Canes Venatici
Ursa Major
30
Izar
Leo Minor
M3
Coma Berenices
20
Arcturus
M64
Leo
Gamma Leonis
M84 + M86
M65
M105
Regulus
10
M87
M66
M95
M96
0
Virgo
ECLIPTIC
Sextans
-10
Spica
M104
Crater
Hydra
-20
R Hydrae
Corvus
M68
Antlia
M83
Hydra
-30
Centaurus
-40
Omega Centauri
13
12
11
Vela
14
10

The Night Sky around Leo
Leo and Virgo are high in
the northern hemisphere
spring-time sky, or the
autumn sky as viewed from
the southern hemisphere.
North of Virgo is Bootes,
whose red giant Arcturus
is the brightest night-time
star in the northern celestial
hemisphere.

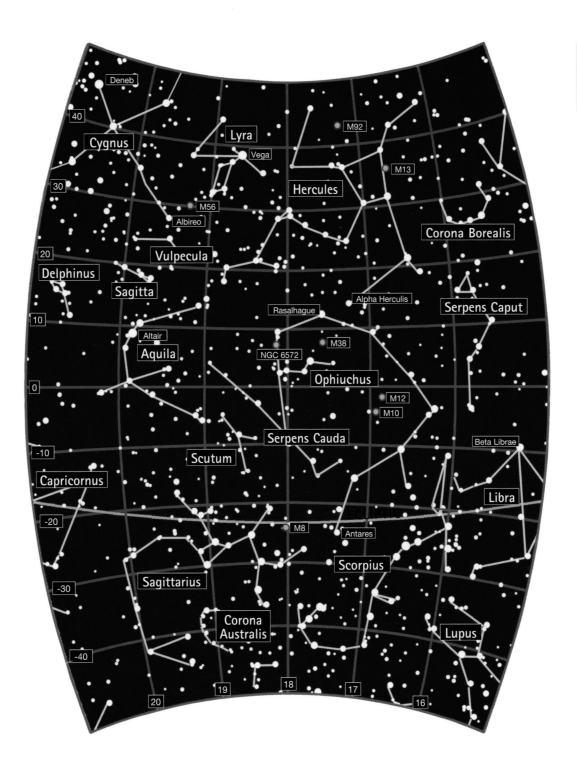

Magnitude scale

0 1 2 3 4 5

Deneb

Cygnus

Lyra

Vega

M92

M13

Hercules

Corona Borealis

M56

Albireo

Vulpecula

Delphinus

Sagitta

Alpha Herculis

Serpens Caput

Rasalhague

Altair

Aquila

NGC 6572

M38

Ophiuchus

M12

M10

Serpens Cauda

Scutum

Beta Librae

Capricornus

Libra

M8

Antares

Scorpius

Sagittarius

Lupus

Corona
Australis

**The Night Sky
around Scorpius**
The core of our Milky
Way galaxy lies in Scorpius
and Sagittarius, high in
the southern hemisphere
winter sky but low in the
summer sky as seen from the
northern hemisphere. Cygnus
and Lyra are the dominant
constellations of the northern
hemisphere summer sky.

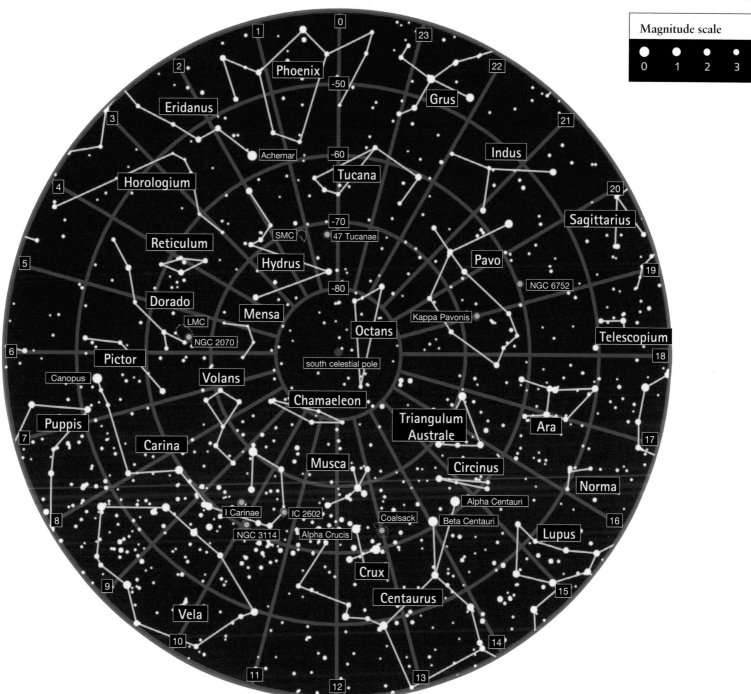

Magnitude scale

0 1 2 3 4 5

The Southern Polar Night Sky

The stars in this chart are always visible throughout the night to observers in southern latitudes such as New Zealand but never visible from Britain. There is no bright star to mark the south celestial pole but the second (Canopus) and third (Alpha Centauri) brightest stars in the night sky are on this chart.

Index

Picture Credits